I0475017

The Nerdy Writer

A Glimpse of the Technical Writer Profession

Donna Fox

ISBN: 146621130X
ISBN 13: 9781466211308

This book covers the opinions of the author based on her experiences and research. These judgments were made independently without payment or expectation of payment from any company mentioned herein. The author is not currently affiliated with any company, product, or services referred to in this book. The purpose of this book is to inform and entertain its readers.

The fact that an organization or website is referred to in this book as a citation or a form of additional information does not mean that the author endorses the organization or website. Further, readers should be aware that the URLs in this book might have been changed or deleted between when this work was written and when it is read.

ABOUT THE AUTHOR

Donna Fox has extensive experience in technical writing and training for both federal and commercial clients. She composes installation and configuration manuals, user guides, online help files, fact sheets, release notes, training plans, eLearning course materials, requirements specifications, news releases, statements of work (SOWs) and web content. These documents support data center, wireless, storage, voice, security, application software, and networking hardware.

During her fifteen-year employment with IBM, she taught numerous network technology courses worldwide. As a sales engineer, Ms. Fox composed customer presentations, demo scripts, and proposal responses on products that integrate voice and data.

Ms. Fox holds a BA in communications from Howard University and two graduate certificates: one in telecommunications from George Washington University, and the second in professional writing from Georgetown University. She also holds a certificate in Leadership from George Mason University.

Ms. Fox welcomes opportunities to mentor young writers and to encourage them to enter the technical writing profession.

Ms. Fox lives in Northern Virginia.

This book is dedicated to tough love. Ethel Fox (Mom) was tough, and Louise Stockett (Nana) was love. I needed them both!

ACKNOWLEDGMENTS

I want to thank the following people for sharing invaluable feedback regarding the draft manuscript of this book. Your comments were extremely helpful.

Dianne Bell
Cheryl Brown
Terry Fox
Ashley Parker
Ted Tanaka
Sidney Underwood

CONTENTS

The Nerdy Writer

A Glimpse of the Technical Writer Profession

Preface

After working in the information technology (IT) field for over twenty-five years, I decided to write *The Nerdy Writer.* It tells the story of how I accepted the fact that I could not plan every step in my life, and, with each unexpected maneuver, came closer to finding my dream profession. Now I want to share what I know.

I decided to write and publish my story because I have written a lot of documents for clients and employers, but none of them are actually mine. As a technical writer, I do not hold the copyright to my work, and I am rarely acknowledged as an author.

Some people say that success begins with a vision or a dream. I did not dream of becoming a technical writer; the profession selected me through a series of chance encounters. When I started this journey, I did not have a specific plan or well thought out goals. I remained open, developed my skills, and adapted to change. I knew I was on the right track because I was blissful. I woke up every morning with great anticipation of the next writing assignment, and I was delighted to learn about the latest technology. I discovered my passion—I am the nerdy writer!

How to Use this Book

This book outlines my professional journey and offers information for individuals who may want to explore the technical writing field. This is not a writing book, but a book about a writing profession. My hope is that it will inspire you to become excited about the field of technical writing.

Think of *The Nerdy Writer* as a roadmap for people who desire to become technical writers. Part One describes how I stumbled into a rewarding career as a technical writer. Part Two furnishes guidance for professionals or students who want to author technology-based documents. This part answers questions and provides direction for people who may want to enter this field.

If you want to read about one technical writer's unconventional journey, start from the beginning (Part One). If you are looking for instructional information about how to navigate within this profession, focus on Part Two.

PART ONE

.

A Writer's Journey

Chapter 1: The Early Days

I am the nerdy writer. No, I did not start out trying to become one; it's been a long journey.

I grew up in a "normal" black family, reared by a single mother and my grandmother Nana, neither of whom had much money. I describe this as normal because the majority of black kids are born into single-parent households. The black family is headed by a single parent; typically a mother who takes on a matriarchal role. My mother was also raised by a single parent. This family structure has been the norm for decades.

Although Mom and Nana rarely complained; we did not have much. My mom, Ethel Fox, was tough as nails, she had to be. She was a tall, slim, and proud woman who was saddled with raising her daughters alone after a bitter divorce. Mom knew we were poor. Foremost on Ethel's mind was to get her daughters out of this pitiful situation.

She knew early on that education was the key to lifting her "disadvantaged" kids from a life of poverty and ignorance into prosperity. She had a cool demeanor but she did not take any mess. Her soft voice delivered powerful words. Her mantra was "Education, education, education!" She wanted her girls to graduate from college, become self-sufficient, and make their mark on society. She believed that education was a kind of springboard that would launch us into attainment. Through education our situation could be altered she thought.

Rockville, Maryland, is a middle-class suburb approximately twenty miles northeast of the nation's capital, the place of my birth. The low-income, public housing units where I grew up was in Rockville's predominately black section. Neighbors worked hard, watched each other's children and tried to improve their circumstances. Everyone was in the same predicament. It was a tight-knit community of loving and persevering people trying to survive.

From my earliest days, oral and written communication was an integral part of my life. I loved colored paper, pens, spiral-bound notebooks, and the large box of Crayola crayons. I used these tools to express my ideas; capture my imagination. When I was not coloring or making collages, I was writing things that made sense only to me; my personal thoughts. If something odd occurred, my Nana would say, with her finger pointed directly at me, "Take note of that, Donna." I guess I took her literally. As I grew older, I began writing journals and poetry.

Curiosity was the bane of my existence. I often asked probing questions. I always inquired, "Why is such and such?" or "How come this and that?" I asked questions all day and into the night. "Donna you are getting on my nerves," Mom warned. I did not know when to ask questions and when to let them go. "Because I said so," was the answer I heard most often. This response led me to ask tougher questions, further fueling my endless inquisitiveness. Back then, you could get slapped for not accepting what you were told. Needless to say, I was always in trouble. Although I was not aware of it at the time, this questioning spirit would serve me well as a technical writer.

I was a bookish kid who loved to read and tell stories. I won several reading contests while in elementary school. Dr. Seuss stories, *The Five Chinese Brothers*, *Babar the Elephant,* and *Madeline* were among my favorites in middle school, I read Ray Bradbury's *The Illustrated Man* and *Fahrenheit 451,* and J. D. Salinger's *Catcher in the Rye.* Reading is fundamental to writing well, and writing every day is essential for great writing. Furthermore, avid readers tend to have a profound understanding of grammar, style, vocabulary, and other aspects of language. This practice accelerated my self-discovery.

Occasionally I read *Ebony* or *Jet* magazines because I enjoyed reading about African Americans' roles in the civil rights movement as well as other issues facing the black community. Both magazines, published by Johnson Publishing Company, encourage advertisers to promote their products to African Americans. The fashion advertisements highlighted models from Ebony Fashion Fair, and the verbiage was positive and self-affirming. These magazines included interviews with black politicians, tributes to celebrities like Diana Ross, and articles on racial injustice. *Jet* and *Ebony* imparted a sense of pride in me and spoke to me in a personal way.

In addition to reading, board games kept me occupied. Monopoly Scrabble, Tic-Tac-Toe, and Checkers were my favorites. Playing games that involved following directions taught me how to focus. We had a set of World Book Encyclopedia (I'm not sure how we afforded it), and we invented a game where we looked up all kinds of nebulous facts. This game helped me to develop cooperative relationships and social skills by teaching me to listen and take turns playmates. Such abilities are useful when collaborating with technologists to compose "help me" or "how-to" documents.

Mom was adamant about exposing her daughters to a broad range of positive experiences. She wanted us to be open-minded and appreciate the "finer" things in life. She always encouraged us to listen to all kinds of music, from jazz to Motown, to eat foods from around the world, and to meet people from all over. We got along well in integrated situations and played with kids from diverse backgrounds.

My mom and Nana took my siblings and me to free concerts and performing arts events around town. I was influenced by these activities. In fact, I performed in several plays. I was the Mad Hatter in *Alice in Wonderland*, and I also had a minor role in *The Sound of Music*. As a teen, I played the lead role in a community play based on a jazz singer's difficult life. Memorizing lines and showing expression on cue resulted in being called "drama queen." In addition to acting, I danced in two troupes; one dance group imitated the Jackson Five, and the other performed modern dance routines. I recited poetry of the Harlem Renaissance whenever I had an audience. I enjoyed being on stage. My family must have been quite shocked when I did not pursue a performing arts career.

Mom never discussed racism or other negative aspects of society. I never knew that I would be denied opportunities. I figured out later that she did not want her children to become jaded. She sheltered us from the harsh realities of inequality. Maybe it's better to look at life through rose-colored glasses than to adopt a defeatist attitude when faced with a ponderous amount of adversity. The negative effects of racism on black youths can stymie ambition. Mom pretended obstacles did not exist.

I was not a stupendous student in high school. I barely remember setting foot in a classroom. Extracurricular activities did not excite me in the slightest, not even the yearbook, school newspaper or book club. I seldom studied, and graduated at the bottom of my class. Still, I knew deep down that I was as smart as my friends and capable of doing outstanding work. Like most high schoolers from the ghetto, I was tough. I thought I could do anything.

Undeterred by my high school performance, Mom continued her relentless drive for me to pursue higher education. "No one is going to take care of you, so you have to be smart and independent," Mom insisted. She constantly underscored the importance of personal responsibility. Her comments had a powerful effect on me. Her words were ingrained in my memory, and they molded my life's journey. But first, I had to rebel against Mom's doctrine.

If I decided to go to college, it would have to be a large, black academic setting that would elevate my level of racial consciousness and expose me to cultural traditions. I needed to associate with African American achievers trying to fulfill the American dream, so attending a historically black college seemed the obvious next step; however, I was not yet sure I was ready.

After high school, I decided to work instead of going to college. Most kids from low-income families wanted to find employment and earn money as soon as possible. I was tired of doing without. I had to find a job in order to obtain a better life and, besides, I wanted to fight the widely held belief that poor people were indigent because they did not work hard enough.

The eighteen-story Parklawn Building was about a mile from my high school and just outside Rockville's city limits. It was a massive structure that housed three federal agencies responsible for public health programs: the National Institute of Mental Health, the Food and Drug

Administration, and the National Health Service Corps, all part of the Department of Health and Human Services (DHHS).

It was relatively easy to land an entry-level position with the federal government back then. I took the civil service exam and scored high enough to become a GS-4 clerk typist at the National Health Service Corps. My high school typing class paid off. Now I was living the good life, earning enough money to rent a room and buy food and new clothes. What more could an eighteen-year-old ask for?

Since I had an outgoing personality, I introduced myself to several clerk typists and secretaries who also worked in the enormous federal building. Many of them shared stories, and I listened intently. I asked questions, and they provided insights about the federal career path. It did not take long to figure out that these women were much older but only one or two levels higher than me. The more I talked to these ladies, the more I doubted my decision not to go to college. I had a strong feeling that this short-term fortune was a dead end. It became clear that being a federal clerk typist would not provide the prosperity mom had talked about. Fortunately, I knew how to get out of this situation, attend college!

In the 1970s, college financial aid was available through a variety of sources. There were grants, student work–study programs, and scholarships. Some financial resources did not require repayment. To access college subsidies, underprivileged smart students only had to be proactive and determined to acquire an education. I was admitted into the prestigious Howard University, firmed up the tuition and board funds, and was ready to embark on the next phase—college life.

Howard University is a private historically black urban institution in northwest Washington DC. Urban universities are influenced by city life and culture; located near restaurants, clubs, and shopping. The old brick buildings are disperse throughout black neighborhoods. There was a connection between these buildings and my hopes and dreams. I asked myself, what will I learn here?

Founders Library sat on the southeast corner of the main campus. It supported the university's academic studies except medicine and law. The high ceilings, extensive reading rooms, and vast stacks of books were on par with white colleges. The walls were oak-paneled giving it a dignified façade. Students and faculty strutted though the long halls with great pride as they perused the wide-ranging resources. This library represented black intellectual history, knowledge, and heritage. Sometimes I fell asleep in Founder's Library, nestled in the stacks, exhausted after surviving an academic challenge.

Undergraduate Credentials

The core curriculum for the undergraduate programs included general studies (English, history, math, and basic science). These courses served as a basis for a broad academic background. I grew up swimming and dancing, so I took these classes to complete my physical education requirements. I trudged through math, science, art fundamentals, and literature courses. Later, I took macroeconomics, microeconomics, and statistics courses that I enjoyed. As I adjusted to the rhythm of college life, I realized that the first few years were providing a foundation.

The more I studied, the more I felt like a nerd. The definition of a nerd is someone who is intelligent but socially awkward. I never attended Howard parties. I wore funky cat-eye glasses and my thick kinky natural hair was untidy. I did pay too much attention to my appearance. I just rolled out of the dorm with mixed-match clothes on headed straight to class. I had developed a few nerdy characteristics.

From time to time, I hung out in the Blackburn University Student Center playing bid whist, a card game that requires talking trash. The Student Center, the newest building on campus, had a television. I watched Eyewitness News on WTOP-TV 9. It had three top-notch black journalists: Max Robinson, Maureen Bunyan, and J. C. Hayward. Max Robinson later became the first African American national nightly ABC news anchor. Max and Maureen were also founders of the National Association of Black Journalists (NABJ); an organization for African American media professionals and journalism students. I was inspired by these prominent journalists. I did not want to be on TV, but I knew I wanted a degree in broadcast journalism.

After completing the basic course requirements, I was admitted into the School of Communications. I declared a major in communications arts, which consisted of courses in print journalism and television broadcasting. During my junior year, I began taking courses that I was really interested in: public relations, graphic communications, voice and diction, mass communications, journalism, copy editing, and television broadcasting. The coursework was rewarding. Learning to report on newsworthy stories was fascinating. I was developing a love for broadcast journalism.

One of my favorite courses was communications law, which dealt with the regulation of television and radio broadcasters. The class covered how and why over-the-air broadcast entities must adhere to federal laws, policies, and procedures. Copyright, profanity, and other content issues were discussed. The class also covered the Fairness Doctrine, the Communications Act of 1934, and the First Amendment with regard to free speech and the press.

Dr. Leroy Miller taught communications law. I am not sure why students called him doctor; he held a doctorate of jurisprudence, but most lawyers don't refer to themselves as doctors. Miller was a scrawny man with a big voice. He could spew out government rules and guidelines for broadcasting. He also had a profound knowledge of the law. Broadcast regulations are straightforward; however, Miller was an analytical thinker and, like most members of the bar, was ready to debate media issues. I loved this thought-provoking course.

10

I would run across campus to arrive early for class with great expectations of Miller's scholarly lectures. Topics debated in Miller's class included the roles of the legal system and mass media play in society, media licensing, the public's right to know, philosophical and practical tensions between government and mass media, freedom of expression versus censorship, invasion of privacy by the media, reputation and defamation, obscenity, and copyright.

While I was in Miller's class, the university applied for a license to launch a television station. The Federal Communications Commission (FCC) required broadcast applicants to ascertain the problems, needs, and interest of their communities prior to obtaining a license. Miller led the ascertainment study for Howard's FCC license. His communications law students accepted the challenge and reached out to Howard's community via a neighborhood-wide telephone survey. I called several hundred people to gather information so that the station could operate in the public's interest. It was a rare opportunity to participate in the development of a television station that would focus on the African American community. The FCC granted Howard a license to broadcast, and Howard University Mass Media (WHMM) was founded in September 1980, three months before I graduated.

The School of Communications' primary goal was to educate young people for professional careers in mass communication. Over the years, the school had built an impressive record of excellent teaching, research, and public service. Through its students, faculty, and staff, the school's print journalism, film, television and radio broadcasting programs were viewed as notable in communications industry.

Being located in Washington, DC was a real plus. There are numerous media sources in the nation's capital including: The Washington Post, the Afro, National Public Radio (NPR), AM and FM radio stations, and several local network television stations. The School of Communications' faculty had professional relationships with these media sources.

An interesting fact, potentially not known by most, is that the founding dean of Howard University's School of Communications was renowned journalist Tony Brown. He is the host of the longest-running series on PBS, *Tony Brown's Journal.* Brown established a standard of excellence for the school's broadcast journalism program.

Most professors in the School of Communications also held positions in the media. Marian Hull, who taught communications technology, worked for the Corporation for Public Broadcasting. Bill Pratt, who taught television broadcasting, was a sports producer and director. Haile Gerima, who taught film classes, was an independent filmmaker. Sam Yette, a tenured journalism professor, was an editor at *Newsweek*. Charles Simmons, another journalism professor, wrote articles for *The Afro American* newspaper. Not only were these outstanding teachers, but they also shared their professional experiences with students. It was clear that the School of Communications would prepare me academically and professionally.

The School of Communications, its faculty, its students, its reputation, and its mission to provide academic and professional instruction for a culturally diverse population of wannabe journalists, film makers, editors, broadcasters and the like. In addition to the academic programs, WHUR-FM 96.3 Howard University's radio station and WHMM-TV, the university's television station was located on campus. Students worked at these on-campus media outlets to gain experience.

College Internships

The faculty encouraged students to take advantage of journalism, public relations, and broadcasting internships. They helped students find these temporary job opportunities. These internships allowed students to receive college credit, gain job experience, and, in some cases, earn extra money.

I surmised that graduating with no communications work experience would not catapult me ahead of other candidates seeking postgraduate employment. To me it's a no-brainer. I had to not only be more qualified than the students at Howard; I had to be more qualified than countless college graduates in DC.

I assumed that employers preferred graduates who had internship experience. Mindful of this, I participated in three internships at local media organizations.

My first internship was in the newsroom at the National Aeronautics and Space Administration (NASA) Headquarters. NASA Headquarters was a short bus ride from Howard's main campus, and the internship paid well. I had two responsibilities at the space agency. First, I perused astronaut applications and crafted biographies for those individuals who were selected for the U.S. astronaut program. Second, I reported aeronautics news that was heard by calling a toll-free number. My sultry but raspy voice announced information about planets, meteors, and other space events. It was an exciting time to work at NASA; the space shuttle program was the major focus of space exploration.

My second internship was at National Public Radio (NPR). Unfortunately, I did not work in NPR's signature news or public affairs programming departments. My job was to collect information about radio transmitter and antennae attrition from numerous NPR stations. After receiving the data, I drafted a report. No it was not exciting, but NPR paid its interns well.

My third internship was the most exciting and entertaining. I was a production intern at a local ABC affiliate, WJLA-TV 7. There was no salary for this internship; I received college credit only.

I worked on a syndicated public affairs TV show titled *The Baxters*. The half-hour show followed a typical family (mother, father, sister, and brother) as it dealt with significant and controversial issues. The first fifteen minutes of the sitcom were produced by Norman Lear in California, and the last fifteen minutes were produced locally in front of a live, preselected studio audience. After the audience watched the show, local on-air talent posed questions that stimulated a hearty debate on the topic presented. I assisted the associate producer with audience selection and writing questions. For example, if the show was about socialized medicine, I made sure that medical students, physicians, Medicare recipients, and others were in the audience. I also wrote promotional copy, greeted the audience, and performed other duties.

While working at WJLA-TV 7, I gained a broad understanding of the internal workings of a television station. There were three departments: news, public affairs, and entertainment. I learned about the role of the floor director, cameraperson, producer, and on-air anchor. I quickly grasped how to use a light meter and read from a teleprompter.

At the conclusion of the internship, the Baxter's host wrote a letter to my professor. I received an A, but the experience was invaluable. Now, filled with renewed enthusiasm about my future, I wondered what the future held for me. How does one know where they will end up, I pondered. What I know for sure at after this internship was that I will become a television writer or journalist.

These three internships gave me a leg up that I would not have otherwise. Attending the School of Communications was vital, but I understood that I could greatly enhance my classroom learning by gaining "real world" experience through journalism-related internships. As a result, I was able list these internships on my first resume.

The school's alumni are active in every aspect of journalism and mass communication. They hold positions with newspapers, magazines, Internet companies, broadcasting companies, and advertising agencies; in public relations, business journalism, and graphic design; and in research, government, education, and industry. There are prominent graduates from Howard's School of Communications.

After graduating from Howard University, I moved back to Rockville, Maryland. Before I know it I settled into my new life or a revised version of my old life. I landed a job as a copy editor at the American Speech Language and Hearing Association (ASHA).

ASHA is a professional membership organization for speech pathologists and audiologists. I edited and formatted manuscripts for two ASHA publications: *The American Journal of Audiology* and *The American Journal of Speech-Language Pathology*. Speech pathologists and audiologists from around the country submitted their scientific reports to ASHA headquarters in hopes of being published in these journals.

There were two deaf women working there, and one of them taught me American Sign Language (ASL).

After working at ASHA for just over a year, I wanted a more exciting job. In the back of my mind, I was still thinking about making my mark in broadcasting. Baffled as to what to do next, I contacted a former college professor, Marian Hull, who worked for the Corporation for Public Broadcasting (CPB). I invited her to lunch hoping that she would help me obtain a position at CPB.

At lunch, the conversation took an unexpected turn. She explained that communications and technology would converge in new and dynamic ways. She strongly suggested that I attend a new graduate telecommunications program being offered at George Washington University (GW). I looked at her like a deer in headlights. I had barely had enough money to pay for undergraduate school, and it was going to take a good portion of my paycheck to pay for lunch. She must have sensed it. She said, "Stop buying people lunch. You cannot afford it. Go to GW. You can get in with your undergraduate GPA, and no graduate record examination is required." I was stunned by her advice. Nevertheless, she was a visionary. I respected her intellect and believed that people would use technology to communicate differently in the future.

I never saw Professor Hull after our luncheon. However, I was certain of one thing that day: I was going to take her advice, and I knew her suggestion would transform my life. It was time for me to take action, so in the spring of 1982, I was accepted into GW's Telecommunications Policy graduate program.

GW's main campus is located in the Foggy Bottom neighborhood of northwest DC. However, courses in this grad program were taught off-campus at locations throughout the metro area. The newly curriculum addressed a broad range of telecommunications concepts and required thirty-six credit hours of coursework. All classes were held in the evening.

The program was multidisciplinary. Engineering courses covered the electronic principles of transmitting voice, data, and video information over distances via cable, micro-wave, satellite, and terrestrial networks. Other courses included management information systems (MIS), which dealt with computer sciences; transaction processing; business applications; data management; information security; and system design. Telecommunications policy and regulation courses were my favorite. These classes addressed the role of government regulators and rules for cable, satellite, and wireless transmission entities. It was a comprehensive program.

While in the graduate program, I met Bob Green, a fellow student and a Telecommunications Manager at Gallaudet University. Gallaudet provides undergraduate and graduate liberal arts education for deaf and hard-of-hearing students. "I know sign language", I told him after class one night. Several months later, he hired me as a Telecommunications Analyst at Gallaudet.

I felt that I met the right person at the right time.

Gallaudet's campus was separated from the community life of northeast DC. An iron fence surrounded the parameter of the campus – segregating the hearing world from the deaf world. American Sign Language not verbal English was used to communicate on campus. Gallaudet had its own culture.

When I joined Gallaudet's staff, a new, campus-wide telephone system was about to be installed. The installation included campus telephones and auxiliary equipment that flashed when a telephone rang. This LED indicator device called a "light relay" was an integral part of deaf culture.

Our first task was to conduct an inventory of the existing telephone equipment, and then write a step-by-step phone switch cutover plan. A cutover is the process of transitioning the legacy telephone system into a new one. Our cutover strategy included planning, managing, and executing configuration tasks for each extension on campus. We spent endless hours configuring telephone extensions, voicemail boxes, telephony features, and long distance services.

The Telecommunication Service Request Form that was used to request, approve, and implement phone moves and changes. It was distributed throughout the campus and used when new hires joined the faculty and staff. In addition, I designed, wrote, and published the campus telephone directory. The Gallaudet Telephone User's Guide described how to use the programmable features. Conducting training classes for faculty and staff was a task I enjoyed immensely. It required that I use my sign language skills to teach deaf learners. I provided technical writing, training, and telecommunications support at a renowned academic institution that relied primarily on non-verbal communication.

In 1982 or 1983, a short time after Bryant Gumbel replaced Tom Brokaw on NBC's *Today* show; he was a guest speaker at Gallaudet. Not sure why he was invited to the university, but he was an eloquent speaker. Still thinking about television broadcasting, I attended his presentation. After he spoke, students asked questions through sign interpreters. One student asked, "How did you make the transition from being a sports announcer to an anchor on *The Today Show?*" I noticed the expression on Bryant's face and said to myself, "Here we go." I expected him to say he went to this school or that school, and predicted he would spout a long list of credentials. I thought he would snidely declare, "We had an actor in the White House." Instead, he coolly made a statement I will never forget: "I have the support from the people who count."

When I began the telecommunications graduate program at GW, I fully expected to walk away with a master's degree. There was no thesis requirement, but students were required to pass a comprehensive exam. I had good grades, so I did not expect to have a problem.

A letter from the assistant dean arrived in May, 1984. It read: "We have received the results of the master's comprehensive examination taken on April 14, 1984, which reports your failure to pass." I had failed my comps. I had spent two years of my life completing thirty-six credit hours and had spent approximately twelve thousand dollars. I did not pass the comprehensive examination I repeated over and over in disbelief. It had three parts, two questions on telecommunications technologies and a question on telecommunications policy. I passed the policy question, but for whatever reason did not pass the question on satellite communications. After hearing the news of my failure, I thought the whole experience was a waste of time and money.

Before I gave up all hope and throw in the towel, I found myself constantly analyzing what could have happened. After all I had been a successful student in undergraduate school. My current grade point average was 3.3; so when I failed the comps, I burned in shame. I remained in agony for quite a while before I mustered the courage to argue my case before the faculty. I requested feedback on my exam. I wanted to know how I could improve. When the dust settled, I concurred with the dean and a professor who was an advocate. They suggested that I take two more courses. I was awarded a graduate certificate in telecommunications the following semester.

At the same time as I was finishing my graduate studies, the telecommunications industry was about to change drastically. AT&T (also known as Ma Bell) agreed to separate from its operating companies in exchange for a chance to enter into the unregulated computer business. In a landmark decision, a federal judge accepted an antitrust settlement from AT&T. The AT&T breakup led to fierce competition in long-distance services and widened new avenues such as data communications for Bell. This series of events also opened the door for unregulated manufacturers to enter into the telecommunications arena. This industry shift was called the divestiture of AT&T.

My post-divestiture plan was to combine my broadcast journalism skills with my telecommunications knowledge and embarked on a prosperous career as a nerdy writer.

Chapter 2: A Career, Not Just a Job

After AT&T's divestiture, I was hired as a technical instructor and course developer in ROLM Corporation's Eastern Region Training Center. ROLM was an early pioneer of digital voice equipment, telephone switches, voice mail systems, and ROLMphones. I taught ROLM's customer engineers how to customize telephone switch and voice messaging software. Each configuration class was three weeks in duration. I taught one class per month, with a week off to catch up on the latest software releases, update course materials based on new standards, and perform other administrative tasks.

After the breakup of AT&T and its Bell operating companies, there was confusion about the changes to the structural and telecommunications boundaries for local and long-distance services. As a result, I developed a course titled "Introduction to Divestiture" that described how new boundaries had been set up between long-distance service providers (e.g., AT&T) and local exchange carriers (e.g., Bell Atlantic).

An understanding of these structural changes was necessary for creating call routing optimization tables is in telephone switches. Call routing tables are used to route telephone calls. These calls are optimized when they traverse the most cost effective or efficient route. Developing a class on the post-divestiture environment was necessary to close a knowledge gap that was essential to understanding telecommunications.

I taught phone system configuration classes for two years before I was promoted to a sales engineering position. A sales engineer (SE) is a technically savvy and detail-oriented salesperson.

ROLM sales engineers were a part of the marketing division, and they were critical to the sales process. They understood the precise technical requirements of product solutions (e.g., digital switching) and used configuration tools to build and price systems for customers. In addition, SEs often accompanied sales representatives on customer calls to ensure that what was sold could be implemented based on the customer's specific requirements. SEs often weighed technical pros and cons. They fully understood compatibility, scalability, performance thresholds, connectivity issues, and interoperability constraints. SEs also wrote customer presentations, responses to requests for proposals (RFPs), and sales collateral.

Soon after I joined ROLM, it was acquired by IBM (also known as Big Blue). At the time, IBM was the largest technology manufacturing company in the computer industry. ROLM, on the other hand, was a much smaller Silicon Valley phone systems provider. The corporate cultures were quite different.

One glaring difference was the IBM uniform. IBM's dress code required all employees (not just customer-facing professionals) to wear dark suits, white shirts, pumps or wing tips, power ties, and very little jewelry. Most professionals understood that a conservative or businesslike appearance set IBMers apart from the competition. Don't judge a book by its cover but the cover can say a lot. Dressing for success tells people you mean business.

What it said about me was that I was willing to give up some of my ethnic identity to play the corporate game. After years of wearing a natural hair style, I relaxed my hair. My colorful prints were exchanged for white starched shirts and black suits. I put my chunky tribal jewelry away to don only on weekends. No, I am not complaining or protesting; I grew accustomed to IBM's corporate uniform. Most days I thought I looked cute.

What I wear says a lot about me as a professional. If I make a presentation wearing jeans an assumption is made. If I make the same presentation in a charcoal-gray suit, a different impression is formed. People make snap judgments. This is human nature. Even though dress codes have been relaxed in recent years, I still know when to pull out the button-down shirt or ruffled blouse to wear with a pin-stripe or Glen plaid suit.

IBM valued its employees. The company spent an enormous amount of money grooming its workers. At one point, up to ten business days of training per year were required. This training consisted of various business courses on how to comprehend corporate financial data such as balance sheets and annual reports. Marketing education included information about building and maintaining client relationships, how to negotiate with customers, how to handle objections, and how to present information eloquently. As a result of this exhaustive training, IBMers were extremely knowledgeable and professional.

In addition to business courses, employees attended technical classes. Technical training was subdivided by IBM product family: mainframes, middleware, software, and network management, for example. I attended networking hardware classes on switches, routers, and multiplexers. I enrolled in courses on networking protocols and technologies (i.e., TCP/IP, frame relay, LAN, WAN, voice, and ATM). At IBM, professionals were allowed to become specialists in specific technical areas. I became a networking specialist.

Like most IBMers, I was a loyal employee and never interviewed outside of the company. The over one hundred year old IBM culture was handed down through hardworking families for generations. The company paid attention to its people. Employees were given a retirement luncheon, a day off for marriage, and flowers as a gesture of condolence at the death of family members. In addition, the IBM Club provided discounted movie, amusement park, and performing arts tickets. Employees felt appreciated. The benefits of working for IBM were extensive. I was dedicated to Big Blue.

The management teams were one of the best aspects of working at IBM. IBM managers attended an MBA-like training session in Armonk, New York. One of my former bosses told me, "Effective management is not just about training; it's about motivating people." Another manager explained, "We were taught how to clear obstacles so that our team members can be successful." Most of the managers I worked for were good listeners and developed two-way communication channels with their team members. They were actively involved with what their people were doing and adept at resolving conflict.

All IBMers were indoctrinated in the company's basic beliefs. These beliefs solidified the corporate culture. One such belief was to strive for excellence. Perfection was not the aim, but rather employees were encouraged to use their talents, abilities, and skills in the best way possible. As a technical writer, I assumed that it meant I should write, rewrite, and edit as many times as required to produce a quality document that exceeds readers' requirements. This was my basic belief.

After the ROLM acquisition, a few ROLM heritage employees were selected to join IBM branch offices. I guess

I fit the IBM "mold" because a year after my sales engineering promotion, an IBM manager offered me a job in the Seattle branch office. IBM wanted to build teams that consisted of both voice and data professionals. I gave this career move ample consideration and discussed it with my family, concluding that it was a worthy professional opportunity. Living on the West Coast would be a different experience.

Job relocations often came with complete moving packages. In those days, it was a widely held belief that you had to move in order to get ahead at IBM. IBMers knew that turning down an IBM opportunity was frowned upon. IBM used to stand for "I've been moved."

Seattle Branch Office

My native Washington DC is a beautiful city with marble monuments, remarkable architecture, historical sites, and the Potomac River. However, nothing prepared me for Seattle's spectacular views. In the distance I could see Mount Rainer and Puget Sound. The downtown cityscape included contemporary skyscrapers, commuters hustling to catch ferries, Pike Place Market buzzing with the scents of fresh fish and flowers, and the world-famous Space Needle. This city was awesome.

In Seattle the air smelled clean and, despite most days being overcast, people are polite and happy. A light drizzle fell for at least nine months a year. Seattle is nicknamed "the emerald city" because of its picturesque landscapes and pristine vistas. Seattle is the home of Starbucks Coffee, Nordstrom's and Microsoft—establishments I've grown to know and love. Seattle has a lot of vitality in the technology arena, performing arts, fine retail shopping, and sporting events.

I was transferred to Seattle to join a team of IBM professionals with extensive mainframe data expertise. I had a solid telephony background. Even though most organizations separated their data centers from their telecommunications departments, IBM marketed integrated voice and data solutions to customers throughout the Pacific Northwest.

After arriving at the IBM branch office, I was informed that I would be working on the Boeing Computer Services (BCS) marketing team. Although Boeing offered computer services to customers, it was best known as an aerospace and defense manufacturer. BCS was located in Bellevue Washington across Lake Washington from Seattle.

The IBM–Boeing relationship was complex: Boeing was a customer, teammate, and competitor. It was one of IBM's huge commercial customers; the IBM team sold products to Boeing. IBM partnered with BCS to bid on federal government solicitations. IBM also partnered with other companies and bid against Boeing.

Boeing Travel managed and coordinated business travel for employees and business partners. Boeing had manufacturing plants and business offices around the Seattle area and offices in Huntsville, Alabama; California; and Ridley Park, Pennsylvania. Boeing Travel distributed an RFP for an Automatic Call Distribution (ACD) system to IBM and other telephone systems vendors. An ACD is a telephone system that distributes or routes incoming calls to a group of agents—in this case, travel agents. These agents booked travel itineraries for Boeing travelers.

The proposed IBM solution included telephones, a switching system, and call routing and reporting software. Since I was the only person on the team who understood telecommunications and had previous experience with this type of telephone system, I led the writing effort for the proposal response. IBM won the bid.

Federal Systems Marketing

After working in Seattle for two and a half years, I moved back east. I transferred to IBM's Federal Systems Marketing (FSM) division in Bethesda, Maryland, a short drive from downtown Washington DC. It felt good to be near family and in familiar surroundings.

Working in FSM was vastly different from working in a commercial branch office. The federal government procurement process is much slower and typically driven by huge RFPs. Government solicitations are massive documents that consist of up to thirteen sections. Each section includes extensive instructions.

The IBM team had a systematic approach to responding to federal proposals. Professionals were assigned to each section. A contract administrator worked on the contract requirements section. Business analysts produced the cost proposal and provided the billing and material information. A marketing representative wrote the cover letter summarizing IBM's qualifications. Finally, sales engineers or technical writers responded to the technical specifications and provided input for the Statement of Work (SOW). It took a diverse team to respond to federal solicitations successfully.

Fully understanding how the government made product and service acquisition decisions was quite challenging. Once potential vendors submitted proposals, it took a long time before contracts were awarded. The sales process took months, occasionally years. Sometimes awards were even protested, which further lengthened the process. There was one positive aspect to this lengthy process and slow pace. After contracts were awarded, the government required assistance from IBMers to deploy its applications or install its hardware. IBMers had adequate time to become proficient its solution and services.

In addition to responding to proposals from agencies, I often wrote demo scripts for proof of concept meetings. On occasion, IBM engineers demonstrated product capabilities in front of key government decision makers. They used demo scripts that documented the customer's current and future environment. During a proof of concept session, demo scripts were used to walk the customer through a series of scenarios that showed how products could be used in various situations. Providing concrete evidence helped customers select IBM as their vendor of choice.

Trade Shows

Trade shows were well attended by federal employees. These shows were free to the public and often drew huge crowds. IBM was an exhibitor and had a massive booth that displayed a broad range of software and hardware products at three trade shows held at the Washington Convention Center (described in Table 2.1). I wrote demo scripts that outlined the presentation steps to demo hardware and software features effectively.

Table 2.1: Trade Shows and Conferences	
ComNet	ComNet was a premier show for inter-networking and communications technologies. IT manufacturers launched new devices and showcased strategic directions.
InterOp	Information technology executives presented papers and provided education on interoperability solutions.
Federal Office Systems Exposition (FOSE)	FOSE focused on IT innovations for the federal government. Free exhibits and executive-level conference presentations were available.

While working at these trade shows, I observed competitors product and service demos. During my breaks, I collected brochures, white papers, and release notes from other equipment manufacturers. This information was useful if competitive analysis documents were required.

The Austin Research Lab

One of the major benefits of working for a multinational corporation is that you can take advantage of a wide range of professional opportunities anywhere in the world. I was offered a short-term documentation project at the IBM research lab in Austin, Texas. This technology lab, one of eight IBM research labs worldwide, was responsible for developing high-speed microprocessors and fast circuit boards for AIX RISK-system desktop workstations. AIX is IBM's version of UNIX.

I spent two months on the sprawling campus of the Austin lab writing content for InfoExplorer, a preinstalled hypertext retrieval library. This tool was used to learn about the RISK-system operating system and it provided an AIX commands reference.

I've Been Moved Again!

Having grown up in urban cities, I was not eager to move to Raleigh, North Carolina. Despite being the state capital, Raleigh felt like I was in the country, not a city. I wanted out of the place as soon as I arrived. The streets in this sleepy town were empty at eight-thirty in the evening, and the city lacked high-caliber cultural activities. I was unfamiliar with the highly attended NCAA basketball tournament fuel by the number of universities in the vicinity. Halfheartedly, I tolerated Raleigh during the week and made the five-hour drive back to DC on the weekends.

There is more land in the Raleigh area than in Washington DC. Homes were not clustered together, and residents were not packed on top of each other in condos or townhouses. It was vastly different from living in the highly populated DC metro area.

I bought a cute, 1500 square foot, one-level ranch home on a third of an acre. My house was in Cary, a yuppie community sandwiched between Raleigh and Durham. The house was white with a bright red door. It sat on a corner lot surrounded by trees. It only costs $126,000. You cannot buy any kind of housing in DC for that amount of money.

I remember one of the locals asking, "Do you know what Cary stands for?" After saying no, I was informed that it meant "Central Area for Relocated Yankees." Shocked after I heard the word "Yankee" but I had to admit over 40,000 professionals were transferred into the area from across the nation. The population almost doubled every year I lived there.

The Research Triangle Park (RTP) consisted of numerous international companies, including IBM. At the time, most companies specialized in technologies or pharmaceuticals. "The Park" or "The Triangle", as folks called it for short, was a North Carolina pseudo–Silicon Valley. Surrounded by three prominent universities; Duke University, North Carolina State University, and the University of North Carolina, the RTP was a highly intellectual environment.

International Technical Support Organization

Although not a welcomed assignment as far as my personal life was concerned, it was an excellent professional decision to make this move to Raleigh. Working at the International Technical Support Organization (ITSO) was like being employed by a technology United Nations. IBM SEs came to the ITSO from different countries. Over time I embraced my transfer.

The RTP ITSO site supported networking hardware, software, and systems management application products. As an ITSO project leader, I was responsible for the 9751 CBX, 9733 Integrated Digital Network eXchange (IDNX) and the 2220 Broadband Switch. These products integrated voice, video, and data onto a high-speed transmission facility (e.g., T1/E1 backbone network).

The ITSO published IBM Redbooks, delivered workshops worldwide, and offered short-term residencies to SEs from other countries. I gained international experience through SEs who had customers worldwide. At the ITSO, I wrote about networking hardware devices that conformed to international standards and specifications.

Residencies

As an ITSO project leader, the focus of my job was three-fold: residencies, Redbooks, and workshops. First, I managed residencies—six- to eight-week assignments for international SEs to come to the ITSO Raleigh to gain new skills with pre-release software or hardware. When there was a new release of 9733 or 2220 code, I set up a lab environment and announced a residency. SEs from around the world who had customers with 9733, 9751 or 2220s volunteered for the residency, and I selected two or three from different geographical areas to participate.

On the first day, my residents were given an outline that provided an overview of the project. This outline not only established expectations, but it also provided instructional guidance. The residents provided feedback and asked questions before they proceeded to the lab. The projects required residents to document installation and configuration steps for the new features or functionality. As the project leader, I made sure they stayed on track.

Redbooks

After six or eight weeks, the residents returned to their home countries with the latest hands-on skills, and I was left with a draft IBM Redbook (a red-jacketed installation and configuration manual). It was sometimes challenging to get the Redbook published in a timely fashion. I faced hard decisions regarding the inclusion of content versus timeliness of information. If a particular feature was not ready to be released, it would not be included in the latest Redbook.

Redbooks were primarily used to assist IBM field engineers with installing and configuring products. However, if the product development groups concurred, Redbooks were often shipped to customers along with IBM products.

Workshops

After a Redbook was published, I developed a workshop that that included instructor and student course materials based on the latest Redbook. Once the workshop was developed, I announced an upcoming workshop tour. Each country replied to the announcement, and I scheduled my tour based on geography and flight routes. I traveled to deliver workshops in cities that I would not otherwise have visited: Sydney, Tokyo, Toronto, and Kuala Lumpur.

Côte d'Azur, too!

In addition to managing residencies in RTP, I also managed them at the ITSO site in La Gaude, located in the south of France. The IBM La Gaude site is on a high cliff roughly fourteen miles northeast of downtown Nice. The squiggly narrow road to the office was only wide enough for 1 ½ cars and was terrifying for some Americans to navigate. French drivers made the perilous trip at high speeds. Once I had arrived at the site and calmed my nerves from the drive up, I could view the French Alps in the distance. This place was magnificent.

A typical workday at the La Gaude site was broken up by a long lunch hour that included delicious food and wine.

After lunch the highly skilled engineers drank coffee or tea around bar-height tables and discuss the latest technology innovations. On sunny days, the engineers took their discussions outside and strolled around the site. The French knew how to enjoy the workday. The workday ended between six and seven o'clock.

Today, the ITSO has changed significantly. Redbooks include not only IBM products but also Cisco and Nortel devices as well as products from other equipment manufacturers (OEM). They have become resources for OEM/IBM interoperability. ITSO project leaders no longer use IBM Book Manager, with its cumbersome tags, as I did; Adobe FrameMaker is currently used to create Redbooks. Residencies are only five weeks or less in duration. I am certain that working at the ITSO is still one of the best IBM jobs.

IBM Education and Training (E&T)

Although the ITSO was a wonderful place to work, I yearned to return to a faster pace and an urban culture. I moved back to the DC metropolitan area and transferred to the IBM Education and Training Division (also known as E&T). As an E&T instructor, I spent most of my time traveling to IBM training centers located in DC; Chicago; San Francisco; Marietta, Georgia; and midtown Manhattan. These were some of my favorite cities. I enjoyed traveling and teaching network technology classes to IBMers, customers, and business partners.

E&T offered training on software, hardware, and middleware products, among other courses. I taught courses on network protocols and specifications, including asynchronous transfer mode (ATM), frame relay, and voice over Internet protocol (VoIP). E&T provided instructor-led classes primarily. However, occasionally customers requested customized classes delivered at their locations. I taught a class in a customer's warehouse in Portland, Oregon.

E&T provided course delivery but the product divisions developed course materials. At times, E&T instructors had to use outdated course materials until course developers created updates that reflected the latest releases and standards. In order to present current information, I updated my classes on the fly to reflect the latest de facto specifications.

A training roadmap was made available to potential students. It listed product courses in the order they should be taken. It was not unusual to have a lecture-only course that introduced concepts and protocols followed by a series of lab-based or hands-on classes. I taught both types of classes.

IBM Career Summary

I realized along the way, that companies that invest in their employees and offer a broad range of work experiences are the best places to establish a career. During my fifteen year career at IBM, I worked for the education and training division, teaching instructor-led courses on networking technologies. I worked in the ITSO and co-authored IBM

Redbooks®, taught technology workshops worldwide and managed 6-8 week projects for SEs to configure networking products in a lab. Early in my career, I was a Sales Engineer in the Federal Systems Marketing division and in a commercial branch in the Pacific Northwest. I composed content for an online help application in a research lab, and I held an international assignment in France.

The company provided a professional foundation, trained me extensively, and gave me opportunities to travel the globe. I welcomed every opportunity that IBM presented to me. I had fifteen good years with the company.

Chapter 3: A Federal Contractor

In the last sales quarter of 1999, IBM shifted its focus and withdrew from the networking hardware business. Subsequently, IBM closed its Networking Hardware Division (NHD) and several hundred network specialists, managers, instructors, and technical writers were laid off. I was devastated when my fifteen-year career with IBM ended. I was knocked down and my dream of retiring from IBM was dashed. It was a huge blow to my identity—I was an IBMer.

I did not know what to do after the layoff, so I did nothing for three months. Finally, I said to myself, "You were groomed professionally by one of the greatest companies in the nation. You have worked with both federal and commercial clients. You were educated in the best universities in DC. You have international experience. What is your problem? You can continue to succeed." After weeks of soul searching, I decided to accept several short-term writing contracts with federal agencies. This chapter covers my government contract job assignments.

The federal government is one of the largest employers in the country. There are a high percentage of government employees relative to the total population. Some are civil servants but there is a surge in federal contractors. After 15 years with IBM, I became a federal contractor. Federal contracting work gave me an opportunity to write documents for various federal agencies.

Technical writers moved from job to job as federal contracts come and go. Each agency had a mission, strategic plan, goals and customer expectations. I benefited from countless professional experiences. These varied writing assignments allow for the development of new skills.

The Supreme Court

Capitol South was the closest metro station to the Supreme Court. When I came out of the station, I was on the House side of Capitol Hill. It is called the House side because the House of Representatives office buildings are located there. One block away was the Library of Congress and the United States Capitol. I walked another block to the Supreme Court, the highest court in the land.

When I passed the white marble pillars of the Court's main entrance each day, I looked up and read the engraved phrase "Equal Justice under Law." Hum, I have some thoughts about that. As a technical writer, I was contracted by the Court to develop user guides, training manuals, course handouts, and other deliverables. I had to put aside my political views; they were not relevant to the job at hand.

During my first day at the Supreme Court, a background check was conducted and photos and fingerprints were taken. All of the contractors attended a security briefing, badges were required to be worn at all times, and nondisclosure agreements were signed. Contractors read the *Law Clerk Manual* and perused the *Visitor's Guide to the Supreme Court* brochure. Some also read various Supreme Court opinions, especially cases that dealt with technologies.

Oh my goodness, here I was, a nerdy writer, walking the same halls as our nation's justices. This experience was impressive, and I was honored to be there.

The contract technical representative (COTR for short) was a delightful woman who provided a tour of the facilities and escorted contractors to a theater to view a film about the Supreme Court. She had worked for the Court for over twenty years. I knew instantly that she would make working there enjoyable and her knowledge would help me develop useful deliverables. She was well organized, and we worked closely together. Based on my study of the Court's opinion writing system and help desk functions, and with a detailed understanding of installed software and hardware,

I was able to develop various relevant documents that met the COTR's requirements.

The Court was undergoing a migration of its outdated desktop applications environment. This was a major upgrade, and the Court's employees required extensive documentation and training. I developed the course curriculum, practice exercises, and handout materials according to specific requirements communicated by the COTR.

The COTR scheduled the software training for the court employees. Technical instructors provided hands-on and instructor-led classroom training for the Court's end-user community. After course completion, a feedback form was used to collect and document comments.

After my contract had ended, the COTR wrote a letter of recommendation. It read: "Donna is a professional technical writer of the highest caliber, who meticulously researches, formats, edits, and proofs her documents. I've received comments from the Court employees who rely on her documentation. Management and personnel in Data Systems and other departments praise her work."

Global Data Dictionary Test Results

When my contract with the Supreme Court ended, I joined an organization that supports federal, state, and local agencies' criminal justice programs. The agencies needed to share information by using leading-edge technologies. This organization was funded by the Office of Justice Programs that awards research and development grants.

At the time, justice and law enforcement agencies procured and deployed information systems and applications independent of each other. These systems did not communicate with each other, and sharing information was challenging. In light of this less-than-ideal situation, a Global Justice XML Data Dictionary (GJXDD) was developed by the Department Of Justice's (DOJ) Bureau of Justice Assistance (BJA), Georgia Tech Research Institute's computer scientists, and the Global XML Structure Task Force (a working group). The GJXDD provided a single repository of criminal justice information.

Agencies and companies began incorporating the data dictionary into their information systems development work; therefore, the GJXDD's performance and scalability had to be tested and verified. A grant to conduct scenario-based performance testing was awarded, a team was established to develop the use cases, and a test plan was crafted. The test team members were graduate students at George Washington University. My job was to coordinate, monitor, and assist the test team members with writing and editing the report before it was given to the DOJ.

At the project's conclusion, a detailed report consisting of recommendations, implementation guidelines, performance tuning measures, and best practices was published and disseminated.

Pension Benefit Guaranty Corporation

Pension Benefit Guaranty Corporation (PBGC) is the federal corporation that protects worker's retirement benefits. This quasi-federal agency had business applications that supported its mission to encourage the continuation of private sector pension plans.

When I began my work there, PBGC was about to rollout an enhanced payment calculation application that would provide increased productivity for staff members and administrators. This release included features such as batch processing, new forms for calculating benefits, drop-down menus, user input validation, support for foreign language reports, and other innovative functionality. In order to communicate the rollout of a new product, I was required to write a user's guide, fact sheet, release note, quick reference guide, and news release. These items were distributed throughout PBGC headquarters and field offices.

Subsequent to the rollout, I edited the graphical user interface (GUI) for consistency and created on-line help files during the software application development process. I reviewed the GUI to ensure that the content was grammatically correct, that formatting was acceptable, and that it was generally easy to use. I checked and rechecked the on-line help tool to ensure that it was useful from a user experience standpoint. My feedback was given to software developers, who updated the code. Once the application was updated, the user experience was clear, concise and simple to use.

State-of-the-art technologies are not usually installed in federal offices. Typically, government workers use legacy applications and tend to work with outdated hardware and software. If you want to work with the latest and greatest technologies, contracting with the federal government may not be the way to go.

Federal Contracting Summary

When you're working on a federal contract, you will most likely have to report to the agency's location. Only in limited situations can federal contractors work remotely. At one agency, contractors had to sign in and out not only when they arrived or left for the day, but at lunchtime as well.

As a federal contractor, you should be prepared to get the worst seat in the house. At one agency, the contractors had to work in a shared office space in the basement. At another agency, the contractors' cubicles were three feet wide, and they had to work shoulder-to-shoulder with their colleagues.

Some federal contracts require the tech writer to compose meeting minutes and save them on a shared drive or upload them to a repository. I know several technical writers who refuse to take meeting notes. My best friend, who is also a tech writer, says, "I hate it, hate it, hate it," when asked to take meeting minutes. I am not a great note taker, but oddly I do not object to taking meeting minutes.

Providers of information technology and services to federal government agencies offered me short-term technical writing, course development, and documentation assignments throughout DC. Contracting opportunities for businesses to sell products and services to federal agencies are numerous. In addition to documentation services, some companies also provided system engineering support, management consulting and professional services to agencies. There are many private firms, called beltway bandits, in the DC metropolitan area encircling the federal government.

Chapter 4: A Commercial Contractor

After eight professionals, including me, were laid off because a contract was over budget, I decided to give federal contracting a break for a while. It never takes long to find technical writing and training services work in Northern Virginia. I narrowed my search to the Dulles Technology Corridor. It is a business cluster of technology companies. This time I looked for a contract position in commercial businesses. In this chapter, I will discuss my writing engagements in the private sector.

Dot-Coms

During the dot-com era, small startup businesses populated technology corridors in most cities. Itchy Networks was one such enterprise. Its mission was to provide technical training and course development services for prominent telecommunications companies. I was hired to develop customized training materials.

I was also hired to teach instructor-led courses on network technologies and protocols, similar to courses I had delivered for IBM E&T. These courses were taught to employees of telecommunications firms.

I earned a lot of money back in the dot-com days. I enjoyed being a technical instructor. Guess it goes back to me being on stage. After the economic market collapsed, the dot-coms were forced to downsize, and they ultimately folded. It was a long time before I experienced that level of prosperity again.

America Online

America Online (AOL) was less than four miles away from my house. I was unfamiliar with AOL at the time, but I soon discovered that it was a prominent, subscriber-based Internet service provider (ISP) that also offered propriety software applications. AOL had merged with Time Warner, a media and entertainment company, in early 2001. I did not make much of this fact; the job would be close to home.

I interviewed with Virginia Brook, the manager of Audience Training and Documentation. She had long, straggly, salt-and-pepper hair and wore faded blue jeans. In fact, everyone I saw in the halls wore jeans. I glanced down at my outfit. I looked out of place in my pinstripe suit.

Virginia explained that she was looking for a technical writer and instructor who could develop instructor and student training materials. In addition, she wanted someone who could deliver instructor-led classes. I showed her my writing portfolio, and she was impressed. She gave me an overview of the project and reiterated that this was only a short-term contract, possibly just six months. She asked if I knew how to use Adobe FrameMaker. I said no, but indicated that I was a quick study.

We met in a high-tech classroom, which had white boards on three walls, jacks and outlets at each seat, and projectors and televisions in the rear. "I would love to teach here," I thought.

AOL rented five buildings on a corporate campus near Washington Dulles airport. From the exterior, the buildings were nondescript multi-story office dwellings. However, the interiors exhibited an industrial design aesthetic similar to New York lofts. The furniture was contemporary, and the decor was modern. It was a cool place to work.

AOL had unveiled a new publishing tool, called the Unified Network Publishing Tool (UNPT). It was used to post promos on AOL.com. I was tasked with developing an instructor manual, student materials, and a post test for the UNPT course. Virginia also asked me to train system analysts on the tool.

While I was working at AOL, the company made plans to create a new and improved AOL.com. A performance quality team was assembled that included engineers, operations professionals, business users, and yours truly. The team focused on the performance and optimization of the web site. AOL was eager to build a top-notch, web-based redesign that equaled or outperformed key competitors, including MSN and Yahoo!

I was also asked to participate in the company's Search Engine Optimization (SEO) Task Force. SEO is based in part on keywords and phrases. How keywords are strategically located and integrated into a website can increase its ranking in a search engine's results. As a technical writer, I played a significant role in facilitating the use of keywords in the website redesign. AOL wanted to compete with Yahoo!, Microsoft, and Google in the search engine arena. Participating in technology task forces and focus groups further enhanced my skills.

Writing for Network Security Engineers

My next short-term technical writing contract position was with a provider of Internet infrastructure services in a network and security operations data center. On my first day, I had to complete heaps of background and security forms. This place was secure. Network monitors lined the walls and hung from the ceiling.

The network and security engineers worked twenty-four hours a day, seven days a week and were tasked with sustaining vital network operations in a multivendor environment.

I assessed the documentation requirements and determined what types of publications were necessary. These professionals needed test plans, troubleshooting guides with detailed network diagrams, and standard operating procedures (SOPs).

In this environment, the SOPs were used to maintain, optimize, and repair networks or subnetworks. These procedures improved communications, provided direction, and were often used for training purposes when new engineers joined the team. When I composed these engineering documents, clarity and accuracy were extremely important. I used short paragraphs and sentences to avoid confusion. All instructions were presented using ordered lists: 1, 2, 3, and so on. I developed concise, task-oriented content so that readers understood how to use and implement leading-edge products, systems, and services. The text was usually simple, interesting, and, most of all, informative. The instructions were often tested to make sure they produced the desired results.

Commercial Contracting Summary

More often than not, technologists do not enjoy writing. Subsequently, companies began hiring technical writers. Federal and state agencies followed suit. At the time, there was a shortage of competent technical writers, so the field was not competitive and finding contracts and jobs was quite easy for qualified technical writers. In the ensuing years, the telecommunications industry went belly-up and so did the dot-coms, and as a result the technical writing profession became more competitive.

Throughout this journey, I have diversified my writing assignments. I have held numerous contract jobs providing technical writing and training services in the public sector. Some of my former clients are quite impressive, and most contracts were enjoyable. Working with network and security engineers to compose standard operating procedures so that they can manage and troubleshoot the internet; was an awesome experience. Developing course materials and training system analysts on a publishing tool to post promos on AOL.com was interesting. By all appearances, my choices of writing assignment have been spot on.

NOTE: *Above all, I learned a great deal from each writing project, and my portfolio reflects a wide range of writing samples. For more information on putting together a writing portfolio, see Chapter 10.*

PART TWO

Notes on the Profession

Chapter 5: Technical Writing Skill Set

Technical writers must possess excellent oral and written communication skills. Moreover, they must be able to conduct research and organize or structure complex information logically. This chapter covers additional competencies (other than writing) that are required of tech writers.

Importance of Audience Analysis

Unlike John Grisham, Dan Brown, and other fiction writers who write for the mass market, technical writers generally compose documents for specific audiences. Audience analysis is the most important step performed by technical writers. In fact, it should be done before a single word is written. Effective communication begins with understanding the target audience. The first questions that should come to mind are, "Who is the audience? Are they specialists or non-specialists? What do they need to know? Who are the readers, and do they understand the subject matter?" Knowing the audience allows technical writers to customize documents to meet the readers' education and skill levels. Technical writers tend to focus on readability and effective use of vocabulary to avoid talking down to readers. It's important to consider the audience. Unfortunately, audience analysis is primarily guesswork!

Some companies provide feedback forms in the back of their manuals or at the conclusion of training classes to determine if the information presented was helpful to the readers or trainees. The readers and students made suggestions and comments. Technical writers evaluate this feedback when composing follow-up materials.

Analytical Abilities and Interpersonal Skills

Analytical skills are typically associated with solving both complex and basic problems. The problem that technical writers are faced with is to review complex information and present it in everyday language. Technical writers are able to gather information, articulate, and make decisions on how this data is presented. Analytical skills are essential.

Technical writers possess specific logical skills. They are energized by working in network operation centers, switch rooms, and development labs; and by writing about servers, telephones, databases and other computer appliances. Tech writers enjoy these environments and tasks. Tech writers have the analytic knowhow to delve into complex information systems, innovative computer solutions, network infrastructures, and other technologies. Technical writers provide details on how a technology works, describe why a particular product development was used, and uncover how a system's architecture approach is beneficial for particular environments, among other topics. The goal of most technical writing assignments is to inform in a coherent manner.

MY STORY—The Analytical Writer
As a technical writer, I was always eager to learn about new technologies. It was my job to ask questions and discover answers. In order to write documents, I need to know how, why, and when. How does this equipment work? How is this solution implemented? How many users are supported? What is required to install it? Under what circumstances does this technology perform at its optimum level?
I posed these questions and translate the responses into user-friendly language for specific readers. I have learned to embrace my relentless curiosity. After all, I document technologies that are constantly evolving. I needed to know how and why.

It Is a Team Sport

Technical writers are required to collaborate with systems engineers, database administrators, network analysts, and sometimes other technical writers effectively. In order for this collaboration to be productive, the technical writer must possess strong interpersonal communications and interviewing skills. Tech writers know how to ask questions tactfully to draw out accurate content from subject-matter experts (SMEs). Some SMEs cannot articulate technical information in a manner that non-technical readers can understand. Obtaining coherent content from SMEs can be like pulling teeth. Working productively with SMEs is critical to the technical writer's document development process. Table 5.1 depicts the interpersonal communications skills required of technical writers.

Table 5.1: Interpersonal Communications Skills For Tech Writers	
Watch	Observe the SME when installing or configuring products or writing code.
Listen	Listen intently—missing one piece of information can change the technical meaning.
Ask questions	Open-ended questions result in detailed explanations; ask for clarification if necessary.
Absorb/learn	Take additional time to process the information; make sure that you fully understand the content. Remember, you are responsible for changing technical verbiage into everyday language, so take good notes.

Contribute	Explain why you are using a list or table to communicate information. Talk about the use of diagrams and other styles or formatting choices.
Follow-up	Coordinate a document review session and incorporate relevant comments and changes.

Tool Gurus

Technical writers are expected to be proficient in authoring tools such as Microsoft Office, Adobe FrameMaker, Author-it, and others. Some organizations will pay for training on these tools, and others will not. If you are serious about being a technical writer, invest in software training and purchase books about specific authoring applications. For more information on tools, see Chapter 9.

Broad Spectrum of Technical Writers

There is a broad range of technical writing expertise. Some tech writers are grammarians and can edit based on structural rules of language. The ability to master and use language is a powerful tool. However, these writers may not have hands-on technical skills; they do not install hardware or software. Many cannot decipher software code, and they rely heavily on information from SMEs.

Occasionally SMEs write the content. The completed drafts are then forwarded to a technical writer to edit and reformat the text based on a style guide. In some cases, they are systems engineers who write—two for the price of one.

At the opposite end of the spectrum are technical writers who possess systems engineering skills. They know how to install and implement networking technologies and other devices. These writers can issue SQL or UNIX commands, and they possess a detailed understanding of databases, biometrics technologies, telephone equipment, and data switches. Hands-on technical skills have become extremely important for writers who develop IT documentation.

Technical writers must constantly embrace new technologies. They must also be familiar with their customers' strategic direction regarding innovations and understand how documentation can best position their customer for the future. Employers want writers to possess some knowledge of what they are writing about, in addition to excellent writing skills. In the IT arena, the skills required are somewhere between writing understanding issues such as connectivity, interoperability, and optimization etc.

I read constantly, perusing trade journals or magazines like Network World, Computerworld, Information Week, and PC World. Reviewing implementation agreements, specifications, and documents from standards bodies (i.e., the Institute of Electrical and Electronics Engineers) is a requirement. There is no other way around it—reading is vital to staying abreast of emerging technologies.

MY STORY—Technical Certifications

As previously mentioned, technical certifications are extremely important. I have five certifications with the following companies: Cisco systems (a network hardware provider), Tellabs (a service provider for enterprise networks), VMware (a virtualization software company), and NetApp (storage systems and software products).

My first certification was a Cisco Certified Network Associate (CCNA), which confirms my ability to configure, install, and troubleshoot Cisco's routers and switches. The CCNA examination validates a basic understanding of voice technologies, routing protocols, and wireless networking. My second certification was a Cisco Certified Design Associate (CCDA). This certification validates an understanding of WAN, LAN, broadband access devices, and Cisco's architecture philosophy of core versus edge products. My third was a Tellabs Sales Engineering certification, which confirms my knowledge of how to sell Ethernet devices, echo cancellers, and smart phones. My forth certification is a NetApp Accredited Sales Associate (NASA) which represents a knowledge of NetApp's products and solutions. My fifth certification is a VMware Certified Professional 5 (Data Center Virtualization). The fact that I hold these certifications signifies that I comprehend technologies and solutions from several manufacturers.

NOTE: *It is easier to write about what I know.*

Basic Coursework May Be Useful

Most people say that they use less than five percent of what they learned in college. I disagree. Some basic courses have been useful—for example, math. In the IT arena, computer devices inter-communicate using binary numbers (ones and zeros). When writing IT documents, it may be necessary to convert values from binary to decimal or hexadecimal. Similarly, network servers, routers, and firewalls may use dotted decimal notation addressing schemes. In order to determine the maximum number of addressable devices on a subnetwork, I had to calculate address values by hand. When a document required pricing information, I estimated telecommunications costs by reviewing tariffs. Basic arithmetic skills are essential.

Although I consider myself artistically challenged, my art fundamentals course provided a few useful skills when incorporating graphics into documents. I am often required to draw diagrams to illustrate networks or create figures that depict hardware configurations. Adding images to manuals clarifies the text. In addition, I captured screens shots of user interfaces to provide step-by-step guidelines with visual examples. Basic design and layout skills are necessary if a graphic artist is not available.

Some of my clients want to be more competitive, gain market share, maximize return on investment, and increase productivity. Business courses provide an understanding of how to compose documents that help make businesses sustainable. Economic fundamentals provide a great source of information for documents that discuss the impact of federal, state, and local regulatory policies. Business and economic principles can be helpful.

Technical Writing Higher Education

People frequently ask me if attending a technical writing academic program is appropriate. I am not against it. If you think that your technical writing can be improved by taking a class or enrolling in a university or commercial training program, please do so. At a minimum, aspiring tech writers will be surrounded by like-minded students and can quite possibly develop mentoring relationships. However, keep in mind that the technical writing profession necessitates other skills that are obtained outside of academia. You will have to acquire skills in collaboration, meeting deadlines, and using authoring tools in the real world.

Most universities and colleges that offer technical writing degrees, certification programs, and courses are part of the English department. High-quality writing requires mastering language, grammar, and syntax fundamentals. Although English teachers provide schooling on style, punctuation, and other basic rules, they may not have experience working in IT or environments that hire tech writers. Effective technical writing takes time, and the ability to communicate complicated concepts requires on-going study.

It was not until the last ten years or so that colleges and universities began offering technical writing degree and certificate programs. The Society of Technical Communication (STC) has a database of college and university academic programs for technical communication. This catalog is in alphabetical order.

MY STORY—Professional Writing Credentials

Almost twenty years after I completed the telecommunications program at GW, I decided to enhance my writing credentials. I was accepted into Georgetown University's graduate certificate program in professional writing. Programs like this one are designed for busy professional writers who seek to continue their education on a part-time or weekend basis. Unlike most writing curriculums that focus on one type of genre, this program exposed students to a variety of writing courses. This program allowed me to continue my writing education without the time commitment associated with undertaking a master's degree.

The program included a diverse group of students; some were recent graduates and a few were older, mid-career professionals like me. I met a grant writer, a journalist, a marketing writer, a speechwriter, and another technical writer. One student was a songwriter. The classes met every Saturday for eight hours. The morning lectures were followed by in-class exercises in the afternoons. There were no exams. The grading system was in place only to provide feedback on writing assignments, which were assigned weekly.

Each week, students were introduced to a professional writer: one of President George H. W. Bush's speechwriters; a marketing director for the National Conservation Center; Tom Clancy's former agent; David Ignatius of the *Washington Post*; Kitty Eisele, writer, director, and producer at NPR; and a travel writer. Although rigorous, it was an outstanding program.

NOTE: *More importantly, I obtained credentials from the most prestigious university in the DC metropolitan area.*

In today's competitive IT market, a college degree is often the minimum requirement for technical writers. Although a degree is not required to become a writer, technical writing is one of those professions where a college education is vital. You do not need a degree from a writing program. You can become a technical writer if you graduate from any college curriculum; however, it may be useful to take courses in computer science, telecommunications, engineering, and other related topics if you plan to write about computer technologies. The earning potential for technical writers who have college degrees is higher than those who do not.

NOTE: *Not all technical writers write about computers. Some write for automotive, aviation, biochemistry, and other industries.*

Summary

Funding educational and training activities to improve both writing and technical skills is an ongoing responsibility for technical writers. Being flexible and having the capacity to adapt to rapidly changing product specifications and standards are essential qualities. Technical writers must embrace new and emerging technologies. They benefit from continuing education courses in databases, networks, hardware, software, and other IT components. This is a profession of endless discovery about how technologies impact our daily lives.

To succeed as a technical writer, you must possess a blend of analytical and interpersonal skills (e.g., team building, project management, and proficiency in mentoring junior writers). You must be skilled at extracting and structuring information from SMEs diplomatically. You must be self-motivated. You may have to meet deadlines—in most cases, without daily supervision.

Technical writing requires a strong command of the English language as well as highly developed writing and organization skills. In addition, working in the field requires excellent skills in using authoring applications. Computer graphics, digital publishing, and photography skills are also useful since many entry-level positions combine these job requirements.

The job market for technical writers is somewhat competitive. Most employers prefer individuals with a bachelor's degree in technical writing, English, journalism, or communications. A graduate degree may be helpful for advancement in the field. There are a number of master's programs available in technical writing.

Practical experience gained on university campuses or during college internships in the IT field are critical elements that can set you above other candidates. Employers seek qualified candidates with relevant experience when making their hiring decisions. Completing internships in information technology environments (e.g., software, hardware, Internet etc.), will increase your chances of getting hired as a full-time employee or receiving a freelance contract.

The demand for technical writers is projected to grow over the next several decades, offering more opportunities and thus reducing competition. In short, the technical writing profession is competitive, but with the right education and skills, a highly dedicated and professional individual will be able to establish a thriving career.

Chapter 6: In the Workplace

In this chapter, I discuss why it is important for companies or organizations to use technical writers. In addition, I will also compare freelancing versus full-time employment. This chapter summarizes how to adopt the same viewpoint as your client or employer, and addresses political challenges often faced by technical writers.

Are Technical Writers Valuable?

Many organizations consider technical writing an important job. Comprehensive product and service documents are required to roll out new technologies in a cost-effective manner. These deliverables are developed for internal professionals and external customers or business partners in order to transfer knowledge. The use and distribution of manuals saves money in training costs, and they can be used to reduce the number of service calls. Tech writer can help organizations be more profitable.

Although technically accurate documents are valuable, few organizations have a full-time tech writer on staff. Usually, a technical writer is hired on a short-term contract basis to assess the requirement, understand the technology, and scribe the deliverable. The temp writer is usually teamed with SMEs, but there are no other writers or editors. After the documents have been developed, they are generally maintained by internal managers. In my experience, finding full-time, long-term writing positions is challenging.

Sometimes technical writers are hired as salaried employees for software development companies, hardware manufacturers and service providers. Some corporations have two or more technical writing job levels: junior technical writer and senior technical writer. If the organization is large enough, there may be a technical writing team. These environments are advantageous for inexperienced writers because best practices can be shared.

When freelance technical writers land assignments, they may not receive paid health benefits or have a career path but are generally paid a competitive hourly rate for the duration of a writing project. Freelancers must know how to do their own taxes or hire an accountant if their taxes are not withheld because they are paid on a 1099 basis. Some freelancers must plan for the lag time between contracts. On the plus side, freelancers can negotiate telecommuting and flexible work hours. Freelance writing can be a great part-time job for people who are balancing other life responsibilities.

NOTE: *For more information on technical writers' pay scale, go to Salary.com and check out other surveys.*

Acknowledgments and Recognition

User guides, administration manuals, and other technical documents seldom include an acknowledgement page. Only rarely is a tech writer acknowledged as an author. In fact, your employer or client can take credit for your writing. They hold the copyright and own the intellectual property.

If you are looking for a career filled with praise and recognition, forget about becoming a technical writer. Oprah Winfrey will not hail you as the most esteemed technical writer of all time. Your talents will not garner critical acclaim. You will not win a Pulitzer Prize. Documenting products and services can be a thankless trade.

If you are employed by someone who has never worked with a technical writer, communicate your value to the organization through status reports and other methods.

What If You Have No Power?

One unwanted work situation to be in as a senior technical writer is one where you have no power or authority. Yes, it's true you were hired because of your extensive writing skills, experience, and credentials. However, you might not make technical writing decisions if you work for a micro-manager who wants to control the entire document development process. In this situation, the manager makes style and formatting decisions, distributes writing assignments, coordinates review meetings, and determines documentation size and content. Micromanagers might meet among themselves, even if it's about documentation. You may ask yourself on a daily basis, "Why am I here?" What should you do if you are being underutilized and undervalued? Hope that you get a new manager or, better yet, a new job! After you land a technical writing job, there may be situations where you will not be given authority over documentation. What's the point? Move on!

You might work with a team of people who declare, "The customer does not read documentation anyway." Is this sentiment relevant? The technical writer is directly or indirectly contracted to compose high-quality, accurate, and concise documents. Whether or not a client or customer reads deliverables is a moot point. There are no excuses for poor-quality deliverables.

A Postscript for Women

Although the statistics are changing rapidly, the IT industry is dominated by men. If you are a female technical writer in this environment, you will be working or interacting with mostly men on a daily basis. If faced with obstacles, pick and choose your battles wisely.

MY STORY— Old Boy's Network
With the exception of one female CIO, the entire executive staff consisted of white men in this organization. No one on the management executive team held an MBA. There are no minorities in upper-level management.
However, this company did a good job of recruiting and hiring women and minorities. Nevertheless, there was absolutely no upward mobility for these employees. Needless to say, turnover was high. Sadly, the executive team either did not comprehend the costs associated with high turnover or they did not care. Fools I would say to myself when I see them gathered among themselves, never acknowledging my presence.

Women must establish solid credentials, be confident in their abilities, exude professional talents, and demonstrate a strong work ethic. Female role models and IT executives are few.

Chapter 7: Related Job Titles

The technical writer job is often confused with abilities and tasks associated with other closely related occupations, such as proposal writer, instructional designer, and documentation specialist. Although potential employers and clients mean well, they are often baffled about which type of writer is appropriate for a particular project. Subsequently, you may arrive at a job interview thinking the hiring manager wants a technical writer when in fact a proposal writer is required. This can be frustrating. Granted, some skills overlap, but a technical writer and a proposal writer are not exactly the same.

This chapter covers the common skills and requirements associated with technical writers, proposal writers, instructional designers, and document specialists (see Table 7.1).

TABLE 7.1—COMMON REQUIREMENTS
• Possess excellent oral and written communications skills
• Meet deadlines
• Format, proofread, and edit documents
• Know how to use Microsoft Office and other authoring tools
• Work well independently and as a team member
• Manage and prioritize multiple documents or projects
• Be detail-oriented
• Collaborate with subject-matter experts
• Ensure that documents are clear, concise, and accurate

What is a Technical Writer?

In addition to the requirements listed in Table 7.1, these are specific tasks associated with technical writing:

- Manage the document development process
- Write, rewrite, edit, organize, and format documents
- Check grammar, spelling, and punctuation
- Prioritize documents as directed by a manager or customer.
- Create diagrams, tables, and templates, and capture screen shots
- Gain technical knowledge through product use, training, and research
- Translate technical jargon for a particular audience
- Upload documents to a content manager for version control

- Conduct team or review meetings
- Adhere to a style guide and other guidelines
- Gather and assess various types of information and effectively put them in writing to covey a specific message

What is a Proposal Writer?

The key difference between technical writing and proposal writing is that the former typically involves "how-to" or "help me" documents. The aim of proposal writing is to sell solutions and services to potential customers. A proposal writer creates responses to requests for proposals or requests for information. The writing is more persuasive; used to convince a potential customer to buy specific services or products. A proposal writer must have excellent writing skills and be able to write in a thorough but influential manner. Proposal writers can be found in almost any industry.

The proposal writer works with SMEs to compose compelling proposal responses. In general, proposal writers write content that informs and persuades potential customers to purchase a solution that meets a requirement or solves a problem. They typically write sales content versus instructional content.

Proposal writers work extremely long hours in a stressful, deadline-oriented environment. Some timelines, set by potential customers, are unreasonable. Since work–life balance often suffers, proposal writers generally earn very high salaries. Finding writers who are willing to give up their free time on a continual basis is difficult.

MY STORY—Proposal Analyst

After a corporate bankruptcy, I was out of work for four months. I reluctantly accepted a job as a proposal analyst with a company that sells, installs, and maintains telecommunications solutions. Proposal writing is not my favorite type of writing.

Two months into the job, my manager informed me that the Air Force was about to release an RFP to procure a biometrics access control and environmental monitoring system. This security monitoring system would be used to arm closets on a base in Delaware. While marketing telecommunications products was the company's core business, my manager had decided to venture into the biometrics technologies arena. There were two other, more senior, proposal writers, so I was a little stunned when he asked me to write the proposal response. We both knew my professional approach to research would help me succeed, so I accepted the challenge.

First, I interviewed our Air Force account executive, who had an extensive relationship with the customer. He provided the sales strategy and, more importantly, knew what the customer really wanted. I wanted to know if he was working with the customer to develop the need, and I asked if he required my help to document the customer's existing environment. It helps to influence the RFP before it hits the streets.

NOTE: *If you help a customer establish a need, you will not have a problem meeting the requirements. If you wait around for the RFP, it's too late—the contract will have been selected.*

Next, I began an exhaustive study of biometrics technologies. I also located SMEs at other offices.

The RFP consisted of thirteen small sections that included an executive summary, management approach, technical overview, and bios for the proposed contract personnel. It was in question-and-answer format. The trick was to craft responses without simply regurgitating the questions.

A small proposal team was assembled: the marketing rep, a contract professional, a business analyst to pull together pricing information, and me, the sole technical writer for this bid. I wrote the entire technical response.

Although I abhor proposal writing, I enjoy winning new business. I take pleasure in knowing that my writing can bring in revenue—and lots of it.

Here is a list of skills required for a proposal writer:

- Write RFP, Statement of Work, and Task Order responses for government bids
- Become knowledgeable about the proposed solution
- Participate in proposal review meetings
- Create requirement compliance matrices
- Adhere to proposal rules, instructions, and deadlines
- Coordinate with a graphic designer (if one is available)

- Maintain a database of past proposal responses. This can be a resource for boilerplate content and past performance information

What is an Instructional Designer?

It's no surprise that technical writing and training frequently go hand in hand: they both involve the knowledge transfer and provide information to specific readers or trainees.

Instructional designers (a fancy name for course developers) generally use a systematic approach to developing training and course materials. They are proficient in determining who, what, when, where, and how training is delivered. They develop instructional materials, from the training plan to post-course assessment, and their goal is to develop an effective educational experience. The instructional designer's unique proficiencies are as follows:

- Conduct training needs assessments; analyze results to determine training requirements and performance gaps
- Compose course objectives and lesson summaries
- Design and develop a training curriculum and course materials using instructional development methods for instructor-led, self-paced, and web-based learning
- Facilitate pre- and post-tests to ensure adequate learning
- Provide train-the-trainer sessions to identify gaps in materials

- Possess strong classroom presentation skills
- Possess skills in e-learning software and authoring tools

Instructional designers typically report to a training manager or project manager. The instructional designer is often required to have a degree in instructional systems design, education, or curriculum development, and the position often requires five or more years of experience in instructional systems design (ISD) methodologies. Master's degrees in education or instructional design are desired as well.

MY STORY—Instructional Designer
I was on contract as a technical writer and course developer for an agency that was about to roll out a software application. Employees throughout the organization required training in order to gain new skills. I gathered information from the agency's users and other SMEs and then employed instructional system design (ISD) methodologies to craft a training plan, design and develop course materials, deliver the class, and assess the effectiveness of the training session.

What is a Documentation Specialist?

The title documentation specialist is often used to refer to a technical writer. A documentation specialist's editing, writing, and authoring tool skills are almost identical to the required abilities for technical writers. However, some documentation specialists work in non-technical industries; for example, health care. They are often in charge of developing and maintaining print or electronic documentation packages such as medical records. The packages are submitted to hospitals as well as regulatory and law enforcement agencies. Strong organizational and planning skills are required because documentation specialists compile or code information from various sources (i.e., patients). They possess knowledge of records management procedures, version control, and report generation.

Unlike technical writers, documentation specialists are often considered administrative assistants in some organizations. The education requirements are lower; some positions require only a high school degree. Documentation specialists possess the following skills:

- Knowledge of business processes and records management
- Basic writing and reading comprehension
- Basic computer proficiency
- Interpersonal communication skills
- Organization and planning skills

Summary

The job title Technical Writer means different things to different people. It may refer to someone who writes proposals or to a professional who develops course materials or compiles document packages. Some skills and tasks overlap with those of a proposal writer, instructional designer, and documentation specialist. Peruse job descriptions to make sure you fully understand and are comfortable with the technical writer's job requirements.

Chapter 8: Technical Document Development Process

When you begin a new job or project by establishing an approach for documentation development, you automatically gain credibility. Some organizations value technical writers and hire them on a regular basis. However, other employers have never worked with technical writers. If you find yourself in the latter situation, it may be wise to establish a technical documentation development process (TDDP). In fact, the technical writer is often asked to implement documentation standards and process improvements. The TDDP will provide a systematic approach to creating clear, concise, and accurate documents.

The TDDP requires that the technical writer is engaged at the beginning of document creation. It also ensures that collaboration is productive. It eliminates duplicate efforts and multiple versions of Word files, and adherence ensures high-quality deliverables. Delivering quality documentation is essential to customer satisfaction. In addition, this repeatable systematic process can be used in any organization.

Steps for Documentation Development

The TDDP steps are listed in Table 8.1:

STEP 1	For new documents, the technical writer and a project manager (in some cases) meet with the customer to assess expectations. The technical writer defines the audience for the document. For existing documents, the technical writer or SME reviews and modifies the old document.
STEP 2	The technical writer distributes writing assignments to SMEs who compose the initial draft (using a template).
STEP 3	The technical writer asks reviewers to review and provide feedback. Reviewers submit feedback in the form of track changes, redlined hardcopy, feedback workflow or via e-mail. Reviewers are important to ensure content accuracy. The more eyes the better.
STEP 4	The technical writer incorporates all changes and forwards the final version to the management team for review.

STEP 5	Once it is approved, the final version is uploaded to the documentation repository (e.g., Microsoft SharePoint or Serena Dimensions). The version control eliminates confusion that can result from the distribution of multiple Word files.

Team Review and Quality Assurance

The technical writer's goal is to provide well-organized documents that adhere to the client's policies and style guide. When a draft document is created, reviewers or SMEs will have an opportunity to read, comment, and recommend improvements. This review provides an invaluable quality check. A number of questions are addressed during this step:

- Does the document meet or exceed the expectations?
- Was the document completed on time and within budget?
- Does this document adhere to policies and the style guide?
- Is this document helpful for the reader?
- Is the content technically accurate?
- Does the technical writer have enough time to incorporate changes and edit?

After the technical writer distributes a peer review request, reviewers, usually SMEs, respond with comments and corrections. It is important that the technical writer incorporate these changes promptly.

Process Barriers

In order for this process to be successful, a few barriers must be considered:

- Lack of adherence to the process will negatively affect outcomes, such as meeting deadlines
- Failure to review documents in a timely fashion will cause delays
- Not engaging the technical writer at the inception of the document creation process reduces the writer's overall project or solution knowledge. (The tech writer must also be invited project kick-off meetings.)
- Copying and pasting content from websites or other copyrighted documents without proper citation is an infringement
- Reviewers of draft documents must provide constructive feedback; otherwise their contribution may be a waste of time
- Check in and check out documents from the configuration manager to maintain version control
- Do not sacrifice technical accuracy for style

Documentation Templates

In some environments, a standard approach for document development is implemented to establish a consistent look and feel. Documentation templates are typically stored and downloaded from a repository (sometimes called a process access library (PAL)). Federal and commercial organizations may have standardized templates for the following types of documents:

- Project and Communications Plan
- System Requirements Document
- System Security Plan
- Quality Assurance Plan
- Risk Management and Test Plan
- Training Plan
- System Design Document
- Deployment and Implementation Plan

As a technical writer, you must familiarize yourself with these templates and the existing formatting and style guidelines. In a federal government environment, standardization generally trumps creativity.

You should report status on the document development process. Write status reports and e-mail them to your client or management team. Make sure that your client or employer can track your progress so that missed deadlines are eliminated. Make sure that all stakeholders can recognize your value.

Content Management Solutions (CMS)

You may be asked to examine how content management can be used provide a centralized approach to develop content, store templates, and auto-generate documents. This endeavor may be prompted by the need to better manage documents so that authors can share, locate, and collaborate more effectively. In older approaches, drafts are developed and saved in an unmanaged decentralized manner. You will review requirements for a content management system (CMS) that will ultimately improve productivity and increase the quality of deliverables. The key drivers might be:

- Reduce repetitive input for project descriptions, executive summaries etc.
- Allow users to retrieve and update content from a central repository automatically
- Control access to data (e.g., view, edit, etc.)
- Improve communications among users
- Make documents more retrievable and re-usable
- Enforce a standard look and feel (i.e. style and formatting)
- Provide a paperless office solution

The core aspect of managing reusable content is determining how the information will be structured. This work may include:

- Defining information types: exec summaries, services descriptions, product prices, logos, diagrams, etc.
- Identifying content rules: font sizes, bulleted lists versus ordered lists, spacing, use of tables/figures, capitalization etc.
- Determine where to create the library
- Decide whether to use folders or documentation sets

- Developing rules for access permissions and management policies

Some sophisticated environments use applications that will automate feedback. This functionality is typically called workflows. The draft document will be sent to a list of individuals for review and approval. Each reviewer will make changes or comments into the draft directly. When all parties have completed their review or approval workflow task, the document is finalized.

Summary

If you are hired by an organization that does not have a formal documentation development process, you may be asked to offer suggestions or craft a process based on your prior experience. When an agreed upon, systematic approach is used, collaboration is more effective and documentation quality is vastly improved. Keep the process simple by minimizing the number of steps, and make sure that all stakeholders buy into the approach. Finally, value team reviews, engage the tech writer at the beginning, adhere to a style guide or template, and incorporate version control via a CMS.

If a documentation development process is already in place. It is your job to fully understand and follow the process and inform your management or client of in-progress drafts under development.

Chapter 9: Technical Writer's Tool Bag

Writers need a tool kit of useful items for crafting documents. Foremost in a writer's tool kit is a laptop, notepad or tablet. These devices must have access to the web and online search engines. A thumb drive, USB drive, or backup drive is handy for backup and recovery.

Microsoft Office Suite

Microsoft Office Suite is the most popular tool used by technical writers. When creating documents, I use Word almost 98 percent of the time. Most of my clients want presentations and course materials created using PowerPoint. Diagrams were drawn using Visio. Project managers often use Excel and Microsoft Project, but I have not been required to use these applications.

I had one client that required that I develop instructor and student materials using Adobe FrameMaker. Other tools for creating documents (e.g., Author-it, Captivate, Acrobat, RoboHelp, or Dreamweaver) may be required. If you're smart, you'll use whatever the client requires.

Resource Materials

Technical writers often have grammar handbooks available or bookmarked in web browsers. These resources explain rules for parts of speech, phrases, sentences and clauses. These guidelines also cover capitalization, spelling and troublesome words. All writing, including technical writing, must have accurate grammar usage to be considered credible. Grammar handbooks are necessary but in the world of blogging, texting, and tweeting some formal rules has become obsolete. The use of outmoded support resources depends on how formal or informal the document is.

NOTE: *The Society of Technical Communications' Baltimore Chapter has a list of jargon to avoid in technical writing.*

Style Guides

Very simply, style guides define standards for style, formatting, organization, design, and the presentation of information. They are used as guidelines. Using a style guide ensures that documents are standardized and abide by specific rules. Your client or employer may look to you for guidance about styles, grammar rules, formatting, and so on. It is imperative that you have a comprehensive understanding of style guides not only to boost your writing but also to increase your value to your client's organization.

There are plenty of style guides available in the public domain, so it is not productive to create one from scratch. If your client or employer has not selected one, here are a few used by technical writers and editors:

- *Microsoft Manual of Style for Technical Publications*
- U.S. Government Printing Office Style Manual
- *Chicago Manual of Style*
- *The Elements of Style* by William Strunk Jr.
- *Style Guide for Business and Technical Communication* by Franklin Covey
- *IBM Style Guide: Conventions for Writers and Editors*

If your customer adheres to a style guide, get a copy of it. For spelling and word choices, I use Dictionary.com and Thesaurus.com. Technical writers often use online search engines to conduct research. Be mindful that no one verifies the accuracy of Internet content, and don't rely too heavily on the results.

Plagiarism Checkers

Plagiarism checkers are typically used by high school teachers or college professors and students to ensure that an author does not copy another's written work without proper citation.

MY STORY—Plagiarism Checker
I never thought I would need a plagiarism checker in my professional career. After being hired by a consulting firm, I was thrown into an unfortunate situation involving SMEs who had decided to borrow content from a vendor's website. The resulting document was forwarded to a client, and the client's staff highlighted each infringement. How embarrassing. This was a regrettable, unprofessional situation. Sadly, I was the only one outraged. As a precaution, I added plagiarism checkers to my desktop. Now I run checks periodically.

Summary

Technical writers are required to be proficient in numerous authoring tools, grammar reference materials, and style guides. The client or employer usually determines which tools and style guide will be used; however, you may be asked to provide suggestions. If you do not have skills in authoring software, your employer will either provide training or reimburse for the cost of classes. However, freelancers are responsible for their training.

Chapter 10: Networking and Landing a Job

There are numerous ways to find a technical writing job. The most important steps are covered in this chapter.

- Embrace professional networking and social media wholeheartedly
- Write a compelling résumé
- Maintain a writing portfolio
- Continue to enhance your professional growth and take personal responsibility for your career development

Professional Writing-Focused Organizations

The most popular membership organization for technical writers is the Society for Technical Communication (STC). This organization offers publications, webinars on numerous topics, chapters in various cities, and a certification program. I recently rejoined STC. For more information, go to www.stc.org.

The National Association of Black Journalists (NABJ) has historically focused on careers and educational opportunities in journalism. This membership organization is now open to anyone who writes content for the Internet and social media. For more information, visit www.nabj.org.

NOTE: *Being a member of a professional organization is an excellent way to increase your knowledge of the field.*

In addition to professional writing organizations, colleges and universities generally have an office of career development (OCD) that provides assistance to students who want to incorporate work experience with academic study. Howard University's OCD helps students in the School of Communications with career counseling, internship opportunities, alumni mentoring, and job placement. Find out if there are mentoring opportunities that can assist you in establishing a career as a writer.

MY STORY—Office of Continuing Education
While I was writing the first drafts of *The Nerdy Writer*, I asked myself, how can I help graduates with bachelor's degrees in journalism or communications enter the technical writing profession? I decided to offer my services and knowledge to Howard University School of Communications' students. After all, I am an alumnus who loved this school.

I contacted the Office of Continuing Education. The director was too busy to take my calls but luckily I met with Jordan Ashford the Placement Services Assistant. I delivered my pitch. "Ms. Ashford, I graduated from this school many years ago. At that time I wanted to become a journalist but through my life's journey I became a technical writer. Technical writing can be a well-paid career and graduates of this school should consider it. I will be more than happy to deliver a presentation to students who are interested in hearing about how to transition from journalism to writing technical documents." Her eyes gleamed. All I need is someone to make contact with the students and a room for the presentation. I will give away free copies of my debut book *The Nerdy Writer* to all attendees, I added. The meeting was brief. However, I left the campus optimistic that I will deliver my mentorship speech to potential mentees.

Several weeks later, I sent a follow up email to Ms. Ashford. Her note read: Your information has been forwarded to the program coordinator for the School of Communications Annenberg Honors Program. They are in need of speakers for next semester classes and you would be a good fit. You should hear from the coordinator soon. I never did.

Using Job Websites and Social Media

Americans are obsessed with two things: technology and social media. Social networking provides a technology-powered, two-way dialogue that embraces the social experience. Social media websites (e.g., LinkedIn, Facebook, Twitter, and Skype) are excellent for gathering referrals, sharing experiences, and building relationships. Users can bring together virtual congregations of professionals with like skills. It is all about making connections virtual or otherwise. Social networking sites will continue to evolve and develop innovative solutions to empower people to establish all kinds of professional associations (e.g. there are several technical writing groups on LinkedIn).

Conduct a personal assessment of your skills. Ask yourself: "What do I want to do? Am I willing to travel? Can I work off-hours and during the weekends?" After gaining a clear understanding of your strengths and limitations, it is up to you to find employment professionals (e.g., recruiters or headhunters) to help in your job search.

When it comes to marketing yourself, don't be passive, and think outside the box. Launch an aggressive personal branding campaign. It may take a little time and energy, but try to create interesting ways to market your writing skills. You can post an "I am a writer looking for a job" video on YouTube. It is a tough job market in some parts of the country, so don't worry about coming off as too gimmicky. You can send broadcast messages to friends in your network, post pictures or links, build pages, and "like" items. You can share your views and ideas worldwide via technology. Networking and blogging with friends, classmates, sororities or fraternities, relatives, recruiters, and headhunters may be helpful. There are few insurmountable obstacles to getting the word out. It is vital that you take advantage of every opportunity to meet new people, so attend social and professional events.

Composing Your Résumé

A résumé is a sales tool. Its purpose is to market your experience and skills in order to obtain an interview. My approach to résumé writing is pretty standard, but it works well, especially when posting on job sites. The goal is to compose an error-free, grammatically correct résumé. Try to limit it to two pages, and, most importantly, keep it up to date.

Most employers are not interested in the potential candidate's objectives, so eliminate an objective statement; opt for a summary statement instead. Employers are more concerned about extensive professional experience, knowledge, and skills. The summary statement is followed by a bulleted list of work experience in chronological order. Each bulleted item begins with an active verb first. Here are a few examples:

- Taught software configuration classes
- Provided technical sales support
- Wrote an education plan

If you have extensive experience, lead with it. If you are a recent graduate, highlight your educational credentials, followed by a list of internships and college writing experiences. It all counts.

NOTE: *For security reasons, I do not include my home address on my résumé. I use my email address and cell phone number.*

Monster, Indeed, and Dice are a few websites that can be used to search for jobs and provide résumé writing tips. Post your résumé on a job search site, and use sites that have a search agent feature. A search agent with "technical writer" as the keywords can be set up to send results to your e-mail address on a daily, weekly, or monthly basis.

MY STORY—Résumé Writer

I worked as a technical writer for a small company that manufactured and sold information security products. This company received congressional funding. When the funding ran out, the company fell under financial distress, and, as a result, filed for bankruptcy protection. After the Chapter 11 filing, the executive staff began negotiating with investors to obtain financing to stay in business. The human resources department did not want employees to quit, but paychecks were not guaranteed. All payments to debtors (e.g., payroll) had to be approved by the bankruptcy court.

During the bankruptcy proceedings, instead of authoring troubleshooting guides, I set up my cubicle as a résumé writing center. I provided résumé writing assistance to forty-four employees. After several months, no funding was found and the company closed. However, most of its displaced employees were ready to enter the job market with new or updated résumés.

Importance of a Writing Portfolio

At the end of an assignment or project, a technical writer has a tangible work product. It is a good idea to collect samples for a writing portfolio. A typical portfolio consists of documents about databases, networking hardware devices, software applications, biometric products, and telephony technologies. It also includes training materials, specifications, user's guides, and operational procedures. It is imperative that you exhibit professionalism, and nothing does it better than arriving at an interview prepared with a writing portfolio. A portfolio shows potential clients and employer's examples of how you structure technical information, and should include writing samples that display a broad range of documentation types. There is no set number of documents or examples that should be in the portfolio as long as you exhibit your best work.

On the other hand, you would be ill-advised to share copyrighted documents, reveal trade secrets, and terms and conditions. For example, if you are on contract with a federal agency that requires a security clearance to document internal processes. These documents may be deemed classified. In general, it is never wise to share proprietary or confidential information.

MY STORY—Writer's Portfolio

A hiring manager once told me, "You are the only candidate with a portfolio." I tried to control the shock on my face. I recall thinking, "Why would someone hire a writer without samples?"

My portfolio includes hard copies of a broad range of documents. I am often reluctant to distribute softcopy samples because they can be distributed anywhere without my knowledge. My portfolio also includes letters of recommendation.

And The Next Job Will Be...

If a technical writer is employed by a large organization, there may be a formal career path in place. A technical writer can be promoted to a senior level. This essentially means that you earn more money and perhaps write about more complex technologies.

There may come a time in your technical writing career where you find yourself not only responsible for your projects or customer deliverables but also the documents from other writers. You may become a team lead or a manager or a senior technical writer who is responsible for the technical writing interns or junior technical writers. Your role may change a bit to include more editing and dealing with management issues. Your leadership and communications skills will be used. The next step in your career path may be management, project management or director of technical documentation and training.

Sometimes tech writers work in quality assurance areas and test systems. They make sure that application user interfaces are clear, concise, and accurate, and that formatting is consistent.

One tech writer went to law school and was later promoted to general counsel. Another acquaintance started as a tech writer and became a federal background investigator. She writes reports based on her interviews with friends of candidates seeking security clearance.

One career step is from technical writer to business analyst. Business analysts identify business needs and determine solutions to business problems. Instead of writing about hardware or software, it is the business analyst's job to develop process improvements, strategic plans, and policy expansions. This job is often related to software development and includes requirements analysis, creating documentation, interviewing experts, and conducting surveys.

MY STORY—Upward Mobility
It is not uncommon for a technical writer to gain extensive experience in a particular technology area and advance into related jobs. I have always been keenly aware that my skills can translate into other professional areas.

Clearly, there are many career paths. What is vital is that you are mindful of your strengths and talents and continue to cultivate new skills. Never lose sight of your true potential, continue to discover your abilities, and market yourself for advancement.

Summary

There are many books about the art or science of technical writing. However, there were no books that offered a real-life account of what happened when a person transitioned from journalism to technical writing. If I were in journalism school today, I would have loved to read a book like this one.

Journalism has changed dramatically in recent years. Anyone with a cell phone thinks they are a journalist. Back in the day, journalism used to be about fact-based, unbiased in-depth reporting. Today it is all about spin and biased commentary drummed up to upsurge ratings. The good news is that investigative skills can be used for technical writing. Print and broadcast journalism is transforming; journalists should consider the technical writing profession.

What I had hoped to achieve in this book is to communicate that I love being a senior technical writer. I wanted to share information about the profession and answer pertinent questions in the process. I sought to inspire you to check out a dynamic, rewarding, and challenging profession. The technical writing field is expected to grow as new technologies are introduced. It is an exciting and competitive field that requires perseverance by those hoping to become successful. As with any field, preparation, hard work, and opportunity are keys to becoming a consummate professional. I am well versed, well read, well-traveled, well informed, and well spoken. In essence, technical writing and training has been my professional life's work. I am the nerdy writer!

Glossary

Acronym	Definition
ABC	American Broadcast Corporation
ACD	Automatic Call Distribution
AOL	America Online
ASHA	American Speech Language and Hearing Association
ASL	American Sign Language
ATM	Asynchronous Transfer Mode
AT&T	American Telephone & Telegraph
BCS	Boeing Computer Services
CCDA	Cisco Certified Network Associate
CCNA	Cisco Certified Design Associate
CMM	Capability Maturity Model
COTR	Contract Technical Representative
CPB	Corporation for Public Broadcasting
DHHS	Department of Health and Human Services
DOJ	Department of Justice
E&T	Education and Training
FCC	Federal Communications Commission
FOSE	Federal Office Systems Expo
FSM	Federal Systems Marketing
GJXDD	Global Justice Xml Data Dictionary
GS	General Schedule
GW	George Washington
IBM	International Business Machines

Acronym	Definition
IDNX	Integrated Digital Network Exchange
ISP	Internet Service Provider
IT	Information Technology
ITSO	International Technical Support Organization
LAN	Local Area Network
LED	Light Emitting Diode
MIS	Management Information Systems
NABJ	National Association of Black Journalists
NASA	National Aeronautics and Space Administration
NCAA	National Collegiate Athletic Association
NHD	Networking Hardware Division
NPR	National Public Radio
OCD	Office of Career Development
OEM	Other Equipment Manufacturer
OJP	Office for Justice Programs
PAL	Process Access Library
PBGC	Pension Benefits Guaranty Corporation
UNPT	Unified Network Publishing Tool
RFI	Request For Information
RFP	Request For Proposal
RTP	Research Triangle Park
SEO	Search Engine Optimization
SME	Subject-Matter Expert

Acronym	Definition
SOP	Standard Operating Procedures
SOW	Statement of Work
STC	Society For Technical Communication
SQA	Software Quality Assurance
SQL	Structure Query Language
TDDP	Technical Documentation Development Plan
VBA	Veteran's Benefit Administration
VOIP	Voice over Internet Protocol
WAN	Wide Area Network
WHMM	Howard Mass Media